POCKET IMAGES

Below
High Hunsley

T0173455

South Cave Market Place between the wars. Walter Scott's shop is now the South Cave News.

POCKET IMAGES

Below
High Hunsley

Malcolm Hall

NONSUCH

Brantingham Church.

First published 1998
This new pocket edition 2005
Images unchanged from first edition

Nonsuch Publishing Limited
The Mill, Brimscombe Port,
Stroud, Gloucestershire, GL5 2QG
www.nonsuch-publishing.com

British Library Cataloguing in Publication Data.
A catalogue record for this book is available from the British Library.

ISBN 1-84588-151-6

Typesetting and origination by Nonsuch Publishing Limited
Printed in Great Britain by Oaklands Book Services Limited

Contents

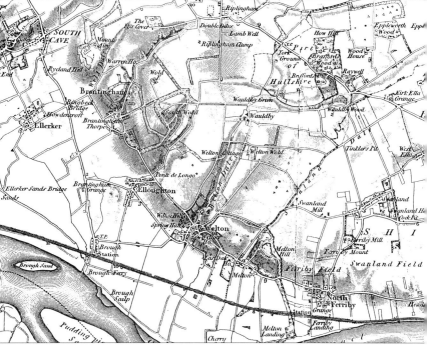

A very early Ordnance Survey map of the region (original scale one inch to the mile) dating from 1862, showing the villages lying between the Wolds and the Humber and with the roads and tracks connecting them, as they existed then. To reach the west from Hull in those days, we passed through most of the settlements, including South Cave, with no sign of an effective trunk road to the west. No wonder the opening of the Hull and Selby Railway in 1840 was the occasion for so much exuberant celebration!

The Area

The Yorkshire Wolds, broad and high between the jutting nose of Flamborough Head and the old town of Malton, become narrower and lower in their more southerly reaches, pausing at the old beacon place of High Hunsley, finally descending at a steeper rate as they approach the northern shore of the River Humber, to leave a low-lying shelf of land between them and the water's edge. This waterside plain, narrow at first, broadens as it rounds the south-west corner of the hills and eventually swells out to join the Vale of York.

To the south and west of these hills, on this riparian plateau, lie a number of communities, some of which have been there so long that their beginnings are hidden in the mists of early time. At least as far back as the Bronze Age, as Edward Wright's work involving the discovery of the Ferriby boats has revealed, men were sailing the Humber and landing on its shores. Some time after that, the Parisi had certainly made the passage from Gaul and were well settled in this region when, during the first century AD, a Roman legion, marching up Ermine Street from its base at Lindum, found itself on the south bank of the wide estuary. There for the time being it halted, leaving the Celts on the far side to their own devices.

In due course, with rebellion in the south and west of the island subdued, the Roman authorities felt able to expand their conquests further and the Ninth Legion was ferried across to the north bank. While the Parisi seem for a while to have remained in occupation of the place we now call North Ferriby, the Romans established themselves a little way upstream, at their fortified settlement which they named Petuaria. From this base, they pushed on northwards and westwards. One of their roads led eventually to Malton. A little way along its line another settlement arose which, by Domesday, had become known as Cave.

As turbulent centuries rolled by, the Romans were replaced by other invaders. The Danes appeared out of the eastern sea, some sailing up the wide-mouthed Humber and on along its tributary rivers, making the conquered lands along their banks part of the Danelaw. Later on still, more changes were wrought, as William and his knights arrived to claim the spoils of war. As the tribal kingdoms of England coalesced into a nation, the ancient settlements became embedded in a local infrastructure which survived and developed, with churches, farms, homesteads, manor houses and the tracks and highways between them becoming established about the land.

On the north bank of the Humber, as elsewhere in the kingdom, the great land-owning families came and went. First, the Conqueror's men held sway: the de Vescys at Ferriby and the Malets at Cave. As these earlier houses fell from grace or died out, others followed. By the eighteenth and nineteenth centuries, it was the city bankers and lawyers, or the merchant princes, raised up on the power of steam and the smelting furnace, who became the new lords of the manor. Names such as Robert Raikes, Henry Broadley, the Barnard family at Cave and Charles Turner of Ferriby came to occupy the stage. By the later nineteenth century, when our photographic record begins, life was still dominated by 'the gentry', which now included some from families which had grown to power and riches through hard work from relatively humble beginnings, perhaps only one or two generations before. Such men as Sir James Reckitt (1833 1924), whose grandfather was a grazier, and Charles Wilson (1833–1907), later 1st Baron Nunburnholme, whose own grandfather began life as a lighterman in the port of Hull.

This is not to forget the ordinary people of the district – artisans and craftsmen, shopkeepers and clerks – whose labours were, to a greater or lesser degree, marshalled and directed by these nineteenth-century entrepreneurs and squires. Here, family names such as the Wilsons and the Gledstones of North Ferriby, or the Waudbys, the Gardhams and the Donkins of South Cave fill the records.

Finally, there came the great social revolution of the present century which, curiously enough, seemed to coincide with a natural dying-out of certain of the erstwhile dominant houses, as they seemed no longer capable of producing heirs. Thus, one by one, the great estates were sold off and broken up into the small parcels of land appropriate to today's social order.

Perusal of the photographs in these pages serves once more to remind us that life in these communities was very different from that in today's world – relatively empty roads, a way of life based largely on the land and villages much more self-sufficient than they are now. Perhaps, after the loss of the village street as a communal meeting place, unthreatened

A deserted South Cave Market Place, with the clock tower of the Town Hall (erected in 1796) reaching high above the neighbouring roofs.

The original, ivy-covered Brough station.

by lethal, fast-moving vehicles, the next most fundamental contrast between nowadays and those earlier times is represented by the change in the character and numbers of village shops. Eighty or ninety years ago, South Cave seems never to have had less than three butchers. Clothing shops too – drapers, as they were called – could always be found, along with grocers, general dealers and the like. Today, such amenities are sadly reduced, if not gone altogether.

And while agriculture formed the major component of activities on land, the broad Humber represented a marine counterpoint of some significance. It was at once a frontier, a workplace and, in those days particularly, a highway too. To the viewer on Ferriby shore, or up on the rising ground at Welton, the river would have appeared a much busier waterway than today's frequently empty expanse suggests. The square sails of Humber Keels populated the estuary, while, from 1815 onwards, 'steam packets' were to be seen plying between Hull and such inland towns as Selby, Thorne and even York. Across their tracks, the ferries ploughed back and forth, on their brief journeys to link the Yorkshire and Lincolnshire shores. The two waterfront villages, Brough and North Ferriby, both had their landing stages, the latter's being used by the local brickworks for the transportation of its products, while that of the former echoed to the feet of countless ferry passengers until that service ceased, some time in the last century.

Then, in 1840, the railway arrived, in the shape of the Hull and Selby Railway Company, its track running arrow-straight across country to Brough, whence it was obliged by the river to bend from its path of rectitude and follow the shore into the city. On 1 July, the first trains were run, in both directions, accompanied by expansive celebrations and general enthusiasm – except, no doubt, in the boardrooms of enterprises such as the Hull Steam Packet Company.

Schooling, too, was different, at least for the majority. Until the passing of the Elementary Education Act of 1870, the dissemination of knowledge to the greater part of the country's children was a random affair, largely dependent on the churches or on wealthy local persons with a sense of social responsibility. Villages such as Ferriby and Cave were no exception and both furnish examples of the practical philanthropy of the Victorian rich, as applied to the provision of at least a modicum of universal education.

With the advance of the twentieth century, the changes began to gather pace, powered by the fires of technology. While the area still housed a largely farming community, together with the great houses and estates of the rich men whose business lay in the bustling city and port of Kingston-upon-Hull, it also began to find industry implanting itself in its very midst. In 1916, men from Leeds came to Brough, land was purchased and a seaplane hangar and slipway were erected by the water's edge. Soon, aeroplanes were regularly to be seen circling the skies. As time passed, these activities were to grow up into Robert Blackburn's aeroplane works – an employer of many local people. In 1937, Capper Pass came up from its base at Bristol and opened its metal refining factory at Melton, while, after the last war, large areas of land disappeared under glass in the enterprise founded by the Bean brothers and still prospering as Humber Growers.

For all that, however, the area today remains predominantly 'country' for, though each community has inevitably been touched and changed by the artificial innovations of the twentieth century, as the housing estates continue to rise and new dual carriageways lay their scars across the land, the fields and hills are never far away.

Malcolm Hall
November 1997

Brough Haven, where a Humber Keel is sharing the moorings with several sailing yachts.

One

North Ferriby

Situated conveniently just a few miles to the west of Kingston upon Hull, North Ferriby village was the first to feel the hand of progress in the eighteenth century. In those days, some of the prosperous merchants and businessmen of Hull felt moved to distance themselves from the city's smoke and odours and to establish new homes in the fresh air of this nearby country village. New houses were built, fit for these well-heeled incomers who, with their servants and carriages, no doubt brought an air of urbanity to this hitherto bucolic community. Ferriby Hall and Ferriby House both date from those times, as did Aston Hall, pulled down in 1970. But, apart from the activities on land, Ferriby folk were well-placed to observe activities on the Humber as it flowed past their doorsteps; in those days the limitations of road communications meant that a flourishing river traffic existed, with packets from Hull – originally sail but changing to steam after 1815 – plying to and from the towns on the Trent and Ouse rivers. North Ferriby possessed its own landing stage, but this seems to have been used largely for the movement of goods, such as the products of the brickworks until the latter place closed down, just after the First World War. When the railway arrived in 1840, North Ferriby was amongst those places provided with a station. In the course of time, this amenity proved to be a further spur to the development of the erstwhile farming village: in the 30 years between 1851 and 1881, the population nearly doubled – from 472 to 837. This rise, however, was dwarfed by the expansion of the next 20 years: by 1901 the inhabitants numbered 4,469.

The Anne Turner cottages, on North Ferriby High Street, built in 1890 and one of a number of examples of that lady's Victorian philanthropy.

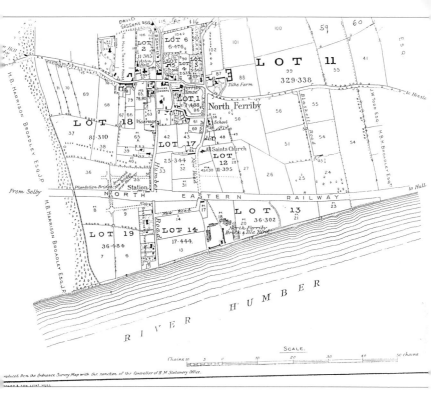

This map dates from 1904, when it was included in the catalogue for the sale of the Turner estate. The Brick & Tile Yard is marked, as are various proposed new roads, such as Marine Avenue and the extended Church Road – witnesses to an expanding North Ferriby. As can be seen, the sale covered a large proportion of the village, including virtually the whole of the shoreline. Charles Turner had died some years earlier and it was on the death of his widow, Anne, that the sale took place, the purchaser being the Hull shipping magnate, Charles Wilson, soon to be created the First Baron Nunburnholme.

Opposite above: North Ferriby High Street, which served as both the main street of the village and the main road from Hull towards the west: in the 1950s, before the construction of the North Ferriby bypass, road travellers between Hull and the south and west found themselves obliged to negotiate the village's High Street, since it was then a part of the A63 trunk road. At some times of day, an uninterrupted flow of vehicles filled the street. Once the bypass had been built, relative peace returned again to the village. On the left can be discerned the gabled roofs of the Anne Turner cottages and the unmistakeable facade of White House Farm

Opposite below: The High Street in an earlier era, looking east.

Next to the Chapel in 1908 stood Giddy's shop. Elijah Giddy died on 4 April 1909, aged only 48. The two bystanders are Harry Pickering, the blacksmith at the Forge, and Tom Holt, a railway porter. If, by the late 1950s, a man had tried to stand out in the road holding a horse's head, he would have been either run over or certified!

The Methodist Chapel on the High Street, which opened in 1878, replacing the previous chapel located in Narrow Lane.

The eighteenth-century Uplands, on the north side of the High Street, with the old Duke of Cumberland pub just visible on the right.

In another view. looking east along the High Street, the entrance to Turner's Lane is on the right, with Ferriby Hall beyond, among the trees.

Left: The old Duke of Cumberland, which no doubt formed one of the centres of social activity down the years. The trio propping up the wall had perhaps been summoned from their proper activities inside, in order to provide the photographer with some human interest.

Below: A Duke of Cumberland has stood on this spot for many years. The earlier building in these photographs was replaced by the present mock-Tudor edifice in about 1930. The splendid open tourer waiting at the kerb was owned by the local physician, Dr Johnson. Was he making a professional call, or just reviving his spirits for a few moments?

The old Duke still survives in this photograph, but only just. Its more modern replacement has been built and stands behind, waiting for its predecessor to be demolished.

The present Duke of Cumberland, with Uplands just beyond it.

Ferriby Hall, one of the country houses built by wealthy men from Hull in the eighteenth century. Occupied by Lord Nunburnholme after he bought up the Ferriby Estate in 1904, it has since the war been occupied by a Club and then by commercial offices.

This group of workers was photographed around 1910, while they were carrying out some work on Ferriby Hall. In the middle of the back row is local builder Harold Kirby.

Above: The War Memorial, at the top of Church Road, diagonally opposite to Ferriby Hall.

Right: The dedication of the War Memorial by the Bishop of Hull on 12 June 1921.

"Lest we forget."

In Honoured Memory of the Officers and Men of this Parish who made the Supreme Sacrifice for their King and Country in the Great War, 1914–1918.

Capt. T. G. Allen,	1st East Yorks. Regt.
Pte. G. Atkin,	2nd Suffolk Regt.
Corpl. G. W. Coates,	4th East Yorks. Regt.
Coy. Sgt-Major G. Goggan,	13th Royal Scots.
Pte. W. S. Everingham,	4th East Yorks. Regt.
Lieut. R. Ferraby,	4th East Yorks. Regt.
Gunner W. A. Ford,	R.N.R., S.S. "Cambric."
Pte. Chas. Grand,	Royal Marine Engineers.
Pte D. Perman Hale,	Royal Army Service Corps.
Lance-Corpl. B. Ireland,	10th East Yorks. Regt.
Pte. Robert Jackson,	10th East Yorks. Regt.
„ G. Johnson,	Durham Light Infantry.
„ C. Lazenby,	Durham Light Infantry.
„ F. Lazenby,	Sherwood Foresters.
„ W. A. Lazenby, M.M.,	11th East Yorks. Regt.
Lieut. N. L. Sissons,	11th East Yorks. Regt.
A.B. Harold Sprowson,	H.M.S. "Bulwark."
Pte. T. Thistleton,	King's Own Y. L. I.
„ A. Thompson,	East Yorks. Regt.
„ John Tomlinson,	Royal Warwick Regt.
„ A. N. Waudby,	7th East Yorks. Regt.
Lce-Corpl. W. A. Waudby,	York. & Lanc. Regt.
Sergt. H. Wilson,	5th Dorset Regt.
„ E. W. Woods,	12th East Yorks. Regt.

Above: Church Road, with the school on the left and the church spire in the distance.

Right: The old school, which was built in 1877 by Anne Turner, dedicated to the memory of her late husband Charles, and which replaced the still older one in Station Road (see page 30). The Turners lived at Ferriby House. Mr Turner was an MP and racehorse trainer who has left us not only Turner's Lane but also, along the Lane, the two-storey building which was his truly palatial stable block and is now the telephone exchange.

Opposite: The list of the fallen of 1914–1918, as it appeared in the Order of Service for the dedication ceremony.

Above: A group of pupils at Ferriby School, in 1904. The headmaster was Mr J.G. Hornby.

Left: A little further along a very rural Church Road, All Saints is fully revealed.

Miss Reckitts Wedding

North Ferriby church has seen many events down the years, both sad and happy. One no doubt in the latter category is that illustrated above. The bride is arriving on the arm of her father, Sir James Reckitt, who lived at Swanland Manor.

Sir James was the third son of Isaac Reckitt who, from humble beginnings, founded the well-known firm of Reckitt & Sons (later Reckitt & Colman) in Hull in 1840, manufacturing and selling starch from its Dansom Lane factory. After early struggles, the company gradually expanded and prospered. By 1917, its total workforce numbered over 5,000. The Reckitts were a Quaker family whose philosophy included a degree of concern for the welfare of their employees. In accordance with this approach, Sir James conceived and built a Garden Village for them near his works in Hull, with well-found houses and pleasant tree-lined avenues.

The bride and bridegroom leaving the church in an open carriage.

The bridesmaids, being similarly conveyed.

The same venue, but a different occasion, as the more sombre dress of the bystanders would indicate.

During the First World War, soldiers of the King's Own Yorkshire Light Infantry were stationed in North Ferriby. The Village Hall, opposite the church, was used as their canteen.

Church Road, probably in the 1930s. The mock-Tudor of the 'new' Duke of Cumberland sees its echo in the facades of the houses and of the row of shops opposite the church. More recently, another change in architectural fashion has seen a very functional brick addition to the lefthand end of the latter, on the corner of New Walk.

CHURCH LANE. N.FERRIBY. E.YORKS.

1440.

Church Road (Church Lane here for some reason), between the church and the railway bridge.

Humber Road: appropriately named, since it led straight from the centre of the village to the bank of the Humber. In 1840, it was cut in two by the opening of the Hull and Selby Railway, the more northerly part later taking its present name of Station Road.

33. 10. RIVER HUMBER. NORTH FERRIBY.

Ferriby foreshore, at the bottom of Humber Road. Beyond the tree lies the area known as Ings Field, used for the grazing of cattle from medieval times, as well as once being the site of the former Ferriby brickyard, which closed down after the First World War.

Today, the area to the left of the picture is a convenient car park for people wishing to enjoy the wildlife site, which has been established just beyond.

On the right is the stretch of shore where, in 1937, Edward and William Wright first discovered traces of the Bronze Age boats, buried in the Humber silt for over 3,000 years (see page 123).

Opposite above: Back up Humber Road, the way is blocked by the railway, which here can be crossed only by the footbridge. Just beyond the lefthand row of houses is the entrance to Marine Avenue.

Opposite below: Marine Avenue at its junction with Humber Road. The children include Marion Ducker, John Hunt and Nancy Sewell in the pram. The gates were closed each night and reopened in the morning.

HUMBER LANE 1424

MARINE AVENUE, N. FERRIBY

Station Road. The building on the right housed the school before it moved to one built in Church Street by Mrs Turner. On the left are Honeysuckle and Moss Cottages, built in the eighteenth century for the servants of Sir Henry Etherington, one-time Mayor of Hull, who also built Ferriby House.

The rear of the old school house on Station Road.

In those far-off days before the upheavals of 1914, Ferriby had its share of that section of society whose way of life was typical of the times but which, with its servants and horse-drawn conveyances, was a world away from today's more egalitarian and machine-driven bustle. This brougham, standing outside Ferriby Garth Cottages, belonged to the Misses Jackson. Their coachman, 'Little Big Dog' Wilson is holding the reins, with Stanley Hotham beside him, both resplendent in their coachman's uniforms and shiny top hats.

The Jacksons lived at Ferriby Garth itself (rear view above), behind the Cottages and facing the High Street at the corner of Turner's Lane. Eliza Jackson died in 1936, at the age of 85. Her younger sister Lucy died in 1952, when she was 97. When she was born, the Crimean War, the Indian Mutiny and the American Civil War had yet to take place; electric lighting was in the future and she had celebrated her fiftieth birthday by the time man first flew at Kitty Hawk. By her death, two world wars had been fought and both the jet engine and the atomic bomb had been developed.

Opposite above: The Jacksons' victoria waits in the Garth stable yard, with young Stanley Hotham on the box seat.

Opposite below: After the War, the times demanded that 'Little Big Dog' Wilson forsake his horses and he found himself at the wheel of a motor car.

Immediate neighbours of the Jacksons were another Ferriby family, the Wrights. Sir William Shaw Wright ran the family oilseed and cattle-cake business in Hull and lived in this house in its own spacious grounds on the north side of Low Street.

It was about 1900 when Sir William apparently decided that more living space was required. Today, he might have opted for a loft conversion, but in those more spacious days he was able to add something considerably more imposing, as can be seen, and the building was henceforth known as Tower House.

Some time before the tower, a new coach house was also erected. Sir William 'kept his carriage', in the shape of the landau on show here, while a second conveyance can be discerned inside. Today, this building survives, as Sands House.

The coach house lay in the spacious lower garden, which lay on the other side of Station Road. With a tennis court set in a leafy precinct and with open fields beyond stretching down to the Humber bank, it must have been a pleasant place to spend some leisure hours.

Above: Sir William and his wife had a number of children, of whom the eldest to survive infancy was Claud, seen here on the left, with his younger brother Horace on the tricycle. In 1889, when the firm of Wright Brothers was going through a difficult period, the two young men (Horace was then 20) went out to South Africa to seek their fortunes. Horace returned two years later.

Left: Florence was the younger sister of Claud and Horace. She is pictured here in the lower garden of Tower House. Behind her can be seen the rear of the old School House in Station Road. Tower House, alas, is another old residence which has been swept away, to make room for several smaller houses more appropriate to present-day needs.

We shall meet Sir Arthur Atkinson again in Chapter 4, since he lived at Elloughton House between the wars. In his retirement, however, he moved to North Ferriby, to live at Ferriby Lodge. He is pictured here in 1954, in his eighties, at the wedding of Peter Hutty and Kathleen Taylor.

Ferriby Lodge c.1959, from the south. Sir Arthur was a director of the Brown & Atkinson shipping company in Hull.

Low Street, with a wall enclosing the upper garden of Tower House on the left.

Low Street in 1908. The boy in the straw hat is Harold Kirby's young son Arthur.

Another view of Low Street, from the other direction, a year or so earlier.

In 1852, the Post Office was located in the High Street, near the Methodist Church and the postmaster was Robert Jackson. By the turn of the century, it had moved to Low Street in the building shown above, which still stands, at the corner of Turner's Lane.

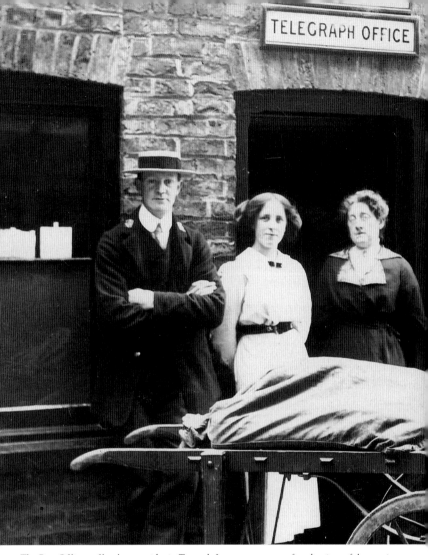

TELEGRAPH OFFICE

The Post Office staff gather outside, in Turner's Lane, some years after the time of the previous picture, a few changes having been made to the outside of the building in the interim while the postmistress is now Miss Fanny Gledstone, having been preceded by her mother and, before that, her father, Mr Robert Gledstone.

From left to right: Mr Woods (postman, killed in the Great War), Pinder (?), Miss Gledstone (postmistress), Hilda Lockey (messenger girl), Mr Mitchell (postman).

By 1929, the Post Office had returned to the High Street, on the corner of Woodgates Lane (see page 44), where it remained for a few years before settling at the corner of New Walk and Church Road.

The Reading Room, that old village institution, was here at No. 5 Low Street until the early 1920s, after which Mackrill's drapers shop occupied the premises for a number of years before the Second World War. Prior to the First World War, the shop was occupied by one William Walker, shoemaker. Standing outside: Miss Hodgeson and a junior assistant.

Opposite above: This group of children were gathered in Low Street, perhaps for a purpose, or perhaps just for the photographer's benefit.

Opposite below: A glimpse of the High Street, from a little way down Turner's Lane.

NORTH FERRIBY 1905

Post Office and Telephone Exchange, Ferriby.

The Post Office, having returned to the High Street, on the corner of Woodgates Lane. It remained there for about ten years, before it was once more on its travels – to Church Road, where it is today.

Woodgates Lane, as it starts to climb up on to the Wold.

This well-known picture of the old smithy, at the entry to an apparently much wider Narrow Lane, is redolent of the atmosphere of the former country village that was North Ferriby. In 1892, the blacksmith was John Caukwell, who died in 1896. After him came Harry Pickering and then George Duggleby, when the former moved to White House Farm. The cottage two doors beyond has seen service as the former Methodist Chapel and then as the Oddfellows Hall.

A last view of a very rural North Ferriby, with White House Farm on the right and, beyond, the Anne Turner cottages. Harry Pickering was the tenant of White House Farm from about 1913, followed by his son Leonard.

Just to the west of White House Farm, the High Street changes into Melton Road, for we are now on our way from North Ferriby to the village of that name. Another change lies in the fact that we have entered Broadley territory. Henry Broadley MP was one of those who came out of Hull to found an estate in the country, first at Melton Hill and then at Welton House. In time, the family's lands became so extensive that it was said that Mr Broadley could ride from Welton to Bridlington, without straying from his own property.

Two

Melton and Welton

Once upon a time, the road used to meander quietly and without interruption from North Ferriby to Melton. No question then of the doubtful pleasure of an exciting negotiation of the dual-carriageway A63 race-track! Today, having reached Melton in one piece, the traveller can at least still continue on the 'Old Road' to Welton without competing any further with the Fangios and Schumachers down on the main road. Welton itself, more than many other places, has retained much of its old charm, thanks perhaps to the way its houses cluster around the church and to each side of the beck, as the latter covers the last stages of its long descent from the Wolds down to the Humber.

The main road through Melton as it was in the early sixties, with Home Farm in the background. Concrete lamp standards have arrived, but the road is free for a lone cyclist, believed to be one Mr Durham.

Laurel Farm, on Melton Old Road. It seems a pity that this handsome house is now largely screened from our eyes by the impenetrable barrier formed by a conifer hedge.

Melton Brickworks, in ruins and now long gone, but still remembered in the name of Brickyard Lane.

This aerial view of Welton village was probably taken in the early sixties: the main road has become dual carriageway, but it is not to the near-motorway standard of today. South of the road lies Pool Bank Farm, the open spaces to the west of which are now occupied by the premises of Humber Growers. Opposite, the white-walled Welton Garth, where many will remember Dr Calvert had his surgery. On the south-western edge of the village, Welton Grange, one-time home of Robert Raikes, stands out, but the other large houses of Welton tend to be hidden by the trees. At the top of the village, the white finger of the estate yard chimney is prominent, behind Parliament Street. Beyond, the empty Wolds disappear into the misty distance.

WELTON VILLAGE E-YORKS
FROM THE CHURCH TOP LOOKING SOUTH

Descending lower to the top of the church tower, the Green comes into focus, with the Green Dragon beyond and Cowgate on the right. Barn House, facing Cowgate, still survives, but the neighbouring houses have long since disappeared.

Down at ground level now and a little further up Cowgate. The mill pond, no doubt, would have been a magnetic attraction for children of many generations.

The Green Dragon Inn. Although much modernised today, the Green Dragon can trace its history at least back to times when it was a posting inn, frequented, it seems, by Dick Turpin.

Dick Turpin's window, they tell us. Whether the celebrated villain did leap from it, as legend has it (straight on to Black Bess, as in the cowboy films perhaps?), must remain conjecture.

Above: A closer view of the background in the preceding photograph, as the village lads disport themselves around Anne Popple's fountain, erected in 1874. Today, the chains and posts have gone and Penrose's shop no longer offers Fry's Chocolate – or anything else come to that as it is now a private house.

Left: A more modern view of the beck, where Cowgate meets Parliament Street.

Above: The Estate Hall, to the right, still survives in the guise of the Memorial Hall, but the building to its left, fronting Cowgate, disappeared many years ago. The shop occupying the righthand end was run by Miss Herring, while that on the left rejoiced in the title of the Welton Pavilion, providing, amongst other things, teas, dinners and stabling for horses.

Right: The old Police House, in Parliament Street, virtually unchanged today. The cottages beyond are also still with us.

53

Perhaps even more than Mackrill's in Ferriby, and from an earlier time, this scene outside Myer's shop in Church Street, with its large staff, is a reminder of what has been lost with the demise of the village shop. The scene is believed to be that of Mr Myers' daughter's wedding. On the left of the shop is the Methodist Chapel. Both buildings are now gone (Creyke House is just outside the photograph, to the right).

Between the wars, Francis Myers had the business, and by the time this picture was taken had invested in this splendid vehicle, being used here for leisure purposes.

Welton House, viewed from the south. This extensive mansion, with its heroically-dimensioned conservatory, was originally built by Robert Raikes in the early nineteenth century. It was then purchased by Henry Broadley MP in 1849 and was the Harrison-Broadleys' seat until 1926. During the war, it housed prisoners of war, fell into considerable disrepair and, regrettably, was completely demolished in 1952, the last Harrison-Broadley having died in 1946.

The Holderness Hunt meeting at Welton House, in the days of its glory.

Three

Brough

Once the Ninth Legion had landed on the north bank of the Humber and the military position had stabilised, the Romans built their fortified town of Petuaria, which became the northern terminus for the ferry from Winteringham. With the retreat of the Roman Empire, its works fell into decay, Petuaria was abandoned and, some time later, the village of Brough grew out of the ruins of its old stones. In the mid-nineteenth century, Alderman T.W. Palmer, Mayor of Hull, the broken column of whose tomb still stands out in Elloughton churchyard, made his home in Brough, at Castle Hill, now demolished. Apart from that, however, Brough largely lay on the sidelines of history for centuries, until, one day in 1916, its modern fate was decided when Mark Swann was sent from Leeds by Robert Blackburn to find a suitable place from which the latter's seaplanes could be launched. The scout reported back, the decision was made, the die was cast. Throughout the twenties and thirties, a succession of Blackburn seaplanes and flying boats descended the slipway behind the hangars, to take off from the Humber's convenient waters.

With its Haven, which once sheltered working keels like the one in the picture, Brough has perhaps greater affinity with the water than its neighbour Ferriby, although it is the aircraft factory and airfield which have come to occupy most of the shoreline, shouldering the village a little way inland.

This aerial view, taken around 1939, provides an excellent overall view of the village and its factory in those days. The aircraft company's hangars line the foreshore, with the seaplane slipway clearly visible. Immediately west is Brough Haven, while the village is broadly enclosed by Saltgrounds Road, Welton Road and Skillings Lane.

Where there's a ferry, there is often a Ferry Inn.

Gaze's butchers shop, at No. 42 Station Road. The delivery van would seem to place this picture in the 1920s.

The dedication of Brough War Memorial.

SKILLINGS LANE, BROUGH · L 9256

Skillings Lane, with not a motor car in sight. The building on the right was once Lodge Farm.

Miss Hunt's shop. This was located just to the right of where Barclays Bank now stands. This point was also the location of the southern toll house of the turnpike from Brough to South Cave, which functioned from 1771 until 1872.

Glenville. Now occupied by Brough Golf Club.

Brough Golf Club in former times. The members are taking their ease outside the 'Monkey House'.

The Station Hotel in the 1880s, when Thomas Kemp was the landlord. It is remarkably little-changed today, apart from, some years ago, having been renamed the 'Buccaneer', to celebrate the last of a long line of Blackburn naval aircraft.

When the Hull and Selby Railway was being planned, the original intention was to route it further north, with a station at Welton. Robert Raikes, who owned the land concerned, was less than keen to see this smelly, dirty, modern contrivance passing through his property and exercised his droit de seigneur. The line was therefore laid further to the south along the Humber shore and Brough got the station (seen above) instead of Welton. The porter in this early picture is Thomas Freeman, later a Parish Councillor.

A later picture of Brough station, in 1895.

The Post Office was at first located on the railway station. Above, a solitary postman stands surrounded by the railway staff for the obligatory photograph.

This photograph was taken in 1904, by which time the original station had been replaced by the one in the photograph, a little further to the east.

It was probably at that time too that the Post Office moved to this building at No. 40 Station Road.

Four

Elloughton, Brantingham and Ellerker

Although Elloughton has long been associated with Brough, administratively and parochially, and the two are now joined invisibly together by the gradual growth of housing developments, its history did not follow the same path as its neighbour. Standing, as it does, more than a mile inland from the waterside, at the foot of Elloughton and Brantingham Dales, its bent will have been more of an agricultural nature (Bulmer's History and Directory of East Yorkshire, 1892, defines its chief crops as wheat, barley, oats and turnips). Both of the smaller villages of Brantingham and Ellerker are fortunate in lying some way distant from the main roads, so that, to some degree, the aspect they retain today is not so far removed from that which they presented at the turn of the century.

Just another hole in the road! Elloughton cross-roads, looking down Main Street towards Brough. Until the building of the new road system, Stockbridge and Welton Low Roads shared with Ferriby High Street the doubtful privilege of constituting part of the main trunk road to Hull.

Looking up Main Street, to the old Half Moon.

Also in Main Street was Cape's butchers shop. Mr Cape was the father-in-law of Mr Gaze, down in Brough.

The original Elloughton garage was run by Mr Swindlehurst.

This party, led by the same Mr Swindlehurst, has just completed the installation of a water treatment works in Elloughton Dale.

Bonnets and boaters! What a pity that no record seems to have survived of the purpose of this large outing of women and children, in these crowded waggons. It certainly seems to have created some interest!

ELLOUGHTON E.YORKS.

Looking from the church, along Church Street towards Main Street, in the early twenties. The house on the left has survived and presents much the same aspect today.

Each summer, a party of the disabled from Hull was brought out to Sir Arthur Atkinson's home at Elloughton House, where they were entertained for the day. Unimpeded by the linguistic niceties of this present age, the event was referred to, without ceremony, as the Cripples' Outing. This long cavalcade of venerable motor cars is bearing some of the guests up Elloughton Dale.

Sir Arthur and Lady Atkinson and their daughters, with their guests.

The guests assembled in the garden during the outing. Whether the English climate always permitted the event to be enjoyed in the open air, with a piano dragged outside, numerous deckchairs and a Punch and Judy show, must be doubted!

The very formal gardens at Brantinghamthorpe. In the 1880s, this house was the home of Christopher Sykes, the younger son of Sir Tatton Sykes of Sledmere. An assiduous courtier, Sykes Minor nearly bankrupted himself in providing lavish hospitality to Edward, Prince of Wales, a frequent visitor to Brantinghamthorpe. Adding self-insult to self-injury, he suffered many indignities at the hands of an insouciant prince. The anecdotes are many: being burnt with Edward's cigar, having brandy poured over his submissive head, all the while enduring the crippling costs of entertaining large house parties to keep his royal visitor amused. One meeting of the Holderness Hunt at Brantinghamthorpe is said to have involved 4,000 on foot, 1,400 on horseback and 1,000 in vehicles. This could not go on: in 1890, the burden was taken over by Arthur Wilson, brother of Charles, who became the prince's host at Tranby Croft for that year's Doncaster race week. He could hardly have imagined that the visit was destined to result in the infamous 'Baccarat Scandal', when an accusation of cheating in a card game, in which Edward was not only one of the players but actually held the bank, was laid against another of his guests. When a civil action for slander in the High Court followed, the newspapers had a field day and the reputation of the heir to Victoria's throne suffered by association.

The road up through Brantingham Dale is still almost as peaceful today.

The Triton Inn at Brantingham was run by Mr and Mrs Watson. Mr Watson earned his living in various activities, including that of a thresher. Some of his equipment lies in the yard in the foreground.

Brantingham village pond.

Brantingham church, nestling amid sylvan surroundings in a fold of the dale.

Instead of proceeding directly from Brantingham to South Cave, we must first call in at the small village of Ellerker.

A leafy corner of the village.

R.F. Glasby, with assistants and numerous barrels.

A corner of Ellerker, photographed in 1908 but which is still instantly recognisable today. Sebastopol Cottage seems to have been so called because the castellated wall in front was built during the Crimean War.

Ellerker Mill, during a pause in its dismantling in 1913.

The remains of the wind shaft, complete with its cross and brake wheel, is caught as it was being dragged off the top of the tower by the traction engine (driven by Sam Downs). Health and Safety inspectors weren't there that day!

Five

North and South Cave

There was a weekly market at South Cave in the twelfth century, which lasted until well into the nineteenth. There was also an annual fair, which involved both livestock trading and entertainments, and which was held over four days during the feast of Holy Trinity. This was still taking place after the First World War. It would seem that the long history of such events – and indeed the existence of the village itself – can be attributed to the latter's location where the old Roman road northwards from the Humber crossing met the road from Hull to the west.

On 10 June 1912, South Cavers came out in force to gaze at the unusual sight of a flooded Market Place.

Two other views of the Market Place at different times and in opposite directions, with the inhabitants going about their business, unimpeded by motor traffic. Although no longer there today, the two footbridges spanned the beck for many years.

Looking south along the Market Place of a quiet country village, with Market Place Farm visible on the left.

Outside Obadiah Martin's – one of Cave's several butchers shops. The baby in the pram, Amy Freeman, was born in 1901.

Bill Smith and his son Bob operated as carriers. The Three Tuns (on the corner of Church Street and Market Place) closed shortly after the First World War. The Cyclists' Touring Club plaque on the wall indicates that the Three Tuns welcomed visitors on bikes.

George Donkin, with his carrier's cart and his not-inconsiderable family, in 1898. George Donkin's cart left for Hull twice a week, every Tuesday and Friday.

Another Donkin (Robert), outside No.13, West End, with what looks like the same cart, but perhaps a few years later. Butler Leake and Martin Gardham, members of two other old South Cave families, are also in the picture.

Church Street leads to South Cave church. It could, however, well have been called Chapel Street: on the right stands the Wesleyan Chapel, now closed, opposite is the chapel-like Church Institute and in the distance, on the left, stands the Primitive Methodist Chapel. Further on still is the building which housed the Girls' School set up in 1841 by Mrs Barnard (see page 90).

Further on down Church Street, by the blacksmith's shop. The stream was put underground in 1966, resulting, one must admit, in a much tidier-looking and perhaps more hygienic arrangement.

This scene would seem to postdate the one above. The stream is still above ground, but the overall impression is a much more attractive one. Perhaps it might have been better if they had left it there?

Above: The original South Cave War Memorial with, behind it, West Lodge. The East and West Lodges were built to mirror the Gothic style of Cave Castle itself.

Right: As today's passer-by can see, South Cave is distinctive in lacking a war memorial consisting of the traditional cross of sacrifice like the one illustrated in these two photographs. It was subsequently destroyed, to be replaced by the present, less-imposing, truncated stone.

Originally, the manor of Cave was split, with two manor houses called East and West Hall respectively. In 1748 Leuyns Boldero, a lawyer from Pontefract, purchased the East Hall estate, later taking the name of Barnard, from another branch of his family. It was his son, Henry Boldero Barnard, who created Cave Castle (above), by the rebuilding of East Hall, and who also acquired the West Hall land. For nearly 200 years, the Barnard family owned the greater part of the parish. The later Barnards appear to some extent to have been examples of that unfortunate phenomenon of those days – absentee landlords. Another member of the family, Charles Leuyns Barnard, fell on the field of Waterloo, serving with the Scots Greys, probably in the charge which shattered the French infantry, but which was then allowed to penetrate too deep into the enemy lines and suffered heavy casualties in the French riposte. With the death of Miss Ursula Barnard in 1938, the line finally died out and the estate, following the general trend, was broken up. The house itself has been a luxury hotel for a number of years.

A typical scene at West End, around the turn of the century.

Woodbine Cottage, Annie Medd Lane, was the home of the said Annie Medd. Some time later it was occupied by Robert Donkin, whom we have already met.

Some of the boys of South Cave School with their master at that time, one James Johnson. The slate held by the boy in the front row proclaims the date to be 21 July 1898. It was the calm before the storm. Within the year, a violent dispute had arisen between the vicar, as chairman of the school managers, and the villagers, with Mrs Barnard from Cave Castle at their head. Some three years before, Mr Barnard, by now deceased, had provided for a new school building, to replace the old schoolroom in the Town Hall. A protracted legal wrangle over ownership of the new school building now broke out. While awaiting the outcome, Mr Johnson, with most of his pupils, retreated to their old premises, where schooling continued. The argument finally ended in defeat for the vicar, victory for Mrs Barnard and the triumphant return of Mr Johnson and his pupils to the new school in 1901.

Four years later, the pupils are assembled on the steps leading up to the new school building. It is a poignant thought that, by 1914, the boys in both this and the previous photograph would have reached manhood – in time for some, at least, to experience the horrors of trench warfare in Flanders.

The girls of Mrs Barnard's School in July 1911.

The Barnard boys' school building is out of sight behind the houses on the right, but it can be assumed that these children constituted some of the pupils.

The road from Beverley, having come down from the Wolds, seen from the Clock Tower as it approaches the roofs of South Cave.

The lady and the child look too well dressed to be involved with the cow which is supposed to be being 'tented' on the verge opposite.

On the slopes of Mount Airy, overlooking South Cave.

Haymaking the traditional way, by Jim Roydhouse and an assistant.

J. Steward, in the bowler, was a seed merchant in South Cave.

These two women were photographed working in the fields near South Cave in about 1916.

George and Arthur Freer, in about 1910.

No, not Heathcliff! We are on the other side of the county from Wuthering Heights and the booted and top-hatted figure is Mr John Cade. The Cades farmed at West End.

Mrs Lundy and Mrs Massey outside Kiln Row, so-called because it was built on the site of an old malting kiln. These cottages were the subject of many complaints concerning overcrowding and smells. They were demolished at the end of the last century and replaced by the Victoria Cottages, still standing at the end of Church Street.

The entrance to the Fox and Coney in pre-motoring days. The figure on the right, sucking on a churchwarden pipe, is a Mr Ward.

It seems that, at some time between the wars, Cave Coal Supplies operated from the Fox and Coney.

North Cave War Memorial, between the wars, with Oldridge's garage in the background.

This photograph is difficult to place in today's much-changed and still-changing Nordham, but is just to the north of Church Street, North Cave.

Church Street, North Cave, with the usual crowd strung fearlessly across an empty road and North Cave church tower rising between the trees. From left to right are: Mabel Towse, Mr Towse, Maud Craven, Daisy Barker, Dorrie Packford, Ollie Packford, Jessie Thornham, Doris Davy, Phyllis Craven, Gyp (the dog). All the houses in this photograph are still standing.

This house, No. 37 Church Street, also survives. Fellmonger is another word for slaughterer.

Right: The camera shutter clicks and Miss Emmy Souter, in her Sunday best and with her new bicycle, is recorded for us to remember her.

Below: South Cave railway station yard. Cave was not blessed with a railway for as long a period as many places. After the opening of the Hull & Selby, several abortive attempts were made to lay another line between the West Riding and Hull, but it was 1885 before one became a reality, with the birth of the Hull & Barnsley Railway. It was not an outstanding success: passenger services came to an end in 1935 and the line finally closed in 1964.

South Cave station, however, was not without its memorable moments. The information accompanying this photograph maintains that it is Edward VII in the car who, having arrived by train, is being conveyed to Brantinghamthorpe. As reported on page 73, Edward, when Prince of Wales, was no stranger to Brantinghamthorpe or the East Riding. The behaviour of the onlookers would certainly imply someone of royal status. If it is Edward, the style of motor car would seem to suggest the early 1900s and that therefore his waiting host would not on this occasion be the hapless Christopher Sykes, who had been forced to sell his estate shortly after the Tranby Croft episode and who had died in 1898.

Scene: the same, as less august personages, in the form of Walter Hunter and Jack Gardham, are being welcomed back from service in the Boer War, which took place between 1899 and 1902. Early British reverses, like the sieges of Mafeking and Ladysmith and other defeats against a numerically inferior but cunning and mobile enemy dismayed the nation. Victory, of a sort, was achieved at length and the population was able to express its feelings of patriotic fervour as its sons returned from the fighting. This scene in South Cave was no doubt typical of many others up and down the land, as the village turns out in force to welcome the two returning soldiers, who have been conveyed by waggon from Brough station.

Two South Cave wedding parties in Edwardian times. The second took place in the well-known Cave family of Waudby.

The South Cave XI in 1909. That year the Australians were in England to contest the Ashes, under their captain, Monty Noble. Possibly, some of the South Cave team managed to get over to Headingley to watch the Third Test, which Australia won by 126 runs. It turned out to be the clincher: England had won the First Test, but at Lords the visitors rose to the occasion and levelled the series. With the Fourth and Fifth Tests drawn, Australia sailed for home having retained the Ashes by 2 matches to one.

Edward VII died on 6 May 1910 and was succeeded by his eldest surviving son. In fact, as Edward's second male heir, the former Prince George grew to manhood with no expection of ascending the throne, but entered on a career in the Royal Navy instead. The death of his elder brother Albert from influenza followed by pneumonia in 1892 changed his destiny at a stroke. Here, the accession of His Majesty King George V is being proclaimed from the steps of the Town Hall on 12 May 1910. The scene is perhaps another example of how the natural forces which propelled the people out of their houses and into social contact on such occasions have been emasculated by the advent of radio and television.

The Coronation of King George V and Queen Mary took place in Westminster Abbey on 22 June 1911. In South Cave, the event was also celebrated, albeit with greater simplicity

Little more than three years later, at the start of the Great War, the scenes were very different. It was to be the first major conflict in which motor transport played a significant part. Here, the Army have descended on the Market Place for the purpose of commandeering what private transport then existed. Round about the same time, in a celebrated episode, the French were doing something similar when they used the Parisian taxis to convey troops to fight in the Battle of the Marne and save the French capital from being overrun in the German advance.

Horses were, to an overwhelming degree, still the army's prime movers, particularly for moving artillery about the battlefield and for the transport of food and ammunition. They, too, were commandeered and in this picture it is the farmers and the carriers who are having to yield to the Army's needs. Thus, it could be said that man's best friend, the horse, faced a compulsory call-up from the start, nearly two years before conscription was introduced for his human comrades.

The crowd at the North Cave Show on 15 July 1909.

The South Cave Subscription Brass Band c. 1925. Needless to say, the band was always an important feature of the proceedings at South Cave Fair.

Six

Industry and Enterprise

By no stretch of the imagination could this area be said to warrant the description of 'industrial'. Nevertheless, in the course of this century, it has been chosen by several companies as the location for their activities, the largest and most long-lasting of which have been the factory and airfield which began life as the Brough premises of the Blackburn Aeroplane Company and which today are still with us, forming one of the major design and production units of British Aerospace.

Later on, other industrial activities made their appearance in the district, although, in those days, road communications with the rest of the country were unlikely to have offered encouragement. However, the proximity of the port of Hull, whose facilities had greatly expanded in the preceding years, would have provided a certain inducement.

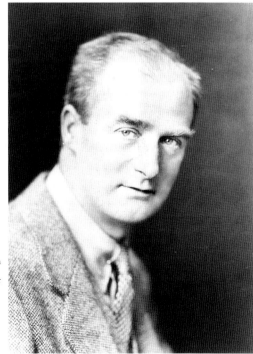

Robert Blackburn (1885–1955) designed his first aeroplane in 1909 and set up his original factory in a disused skating rink in Leeds. Within a few years, either by fate or by design, a large part of its output came to be represented by seaplanes. So it was that, in 1916, Blackburn chose Brough as a base from which the company's seaplanes (and also landplanes) could be test-flown.

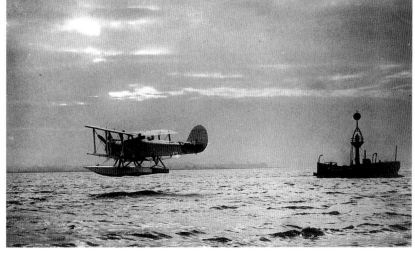

Almost from the beginning, Blackburns found a special affinity with marine aviation. It was natural, therefore, that between the wars, during the heyday of the big flying boat, they should design and build a number of these galleons of the air. Their size and the relatively small demand for them meant that only ten were built in all, but the sight of these stately craft taking off and landing must have added a new dimension of interest to watchers on both shores of the Humber in those days. The first of the breed was the Iris, four of which went into service with the RAF's 209 Squadron. It was followed by the Sydney, of which only one was built, and finally the Perth (above) which replaced the Irises.

Opposite above: The hangars and airfield at Brough, in the early days. Events were later to decree that this small Humberside village would become, by the end of the 1920s, the centre for all Blackburns' work, not only the flying, but the design and production as well.

Opposite below: During the First World War, Blackburns largely built other companies' aircraft by sub-contract, expanding its own workshops and experience thereby. After the war, the Dart seaplane was a successful early example of the company's own designs. A version of the Dart was the Velos (above), which was sold to the Greek government.

The three Blackburn brothers, Charlie, Robert and Harold, are here flanked by two of Robert's earliest employees – Mark Swann on the left and George Watson on the right. The former is generally credited with the discovery of Brough in 1916 as the potential new Blackburn base, having been sent on reconnaissance for that purpose. George Watson was one of Robert Blackburn's first mechanics. He left the firm to spend the war flying with the RNAS, mostly in the Mediterranean, then returned to Blackburns, finally retiring as Experimental Manager after more than forty years with the company.

Opposite above: As the thirties advanced and war loomed once more, Blackburns built its first military monoplane since the very early days. The first prototype Skua (above) made its maiden flight 9 February 1937.

Opposite below: Later on in the Second World War, the company spent a considerable time developing the Firebrand strike aircraft. This aircraft came to grief in a corner of the airfield, in front of the Flying Club.

The first NA39, XK486, was carried by road transport to the Royal Aircraft Establishment airfield at Bedford, where it made its first flight 30 April 1958. All subsequent aircraft, right up until production ceased, were towed from the factory at Brough to the company's flight test airfield at Holme-on-Spalding Moor – a distance of 15 miles. Naturally, for this exercise, the folding wings were a useful feature, although this particular advantage was perhaps not foreseen by my Lords of the Admiralty when they drafted the original specification for the aircraft. However, when, over a number of years, you are required to move more than 200 aircraft from A to B in this manner, it is scarcely to be hoped that all in your world will remain forever serene. Buccaneer Mk.2 XT270 was destined to be the aircraft which would see a minor hiccup in what had become an efficient and routine operation. The aircraft itself is no doubt in good order, but its erstwhile slave, assigned to convey it to its appointment with the skies, appears to have run out of puff, requiring the dispatch of a second one, to assist or take its place.

Opposite above: Blackburns was a major local employer for many years. Its heir, British Aerospace, still is. Here some of the employees are seen queueing at the clocking stations.

Opposite below: On 20 June 1950, local eyes were popping as a new and imposing shape appeared in the Humberside skies. The Universal Freighter was the prototype of the aircraft which was to become known as the Blackburn Beverley.

Twentieth-century technology meets man's eternal food source. A total of 209 Buccaneers were built, going into service with the Royal Navy, the Royal Air Force and the South African Air Force. It was in 1963, while the flight test development of the early Buccaneers had still some years to run, that the name of Blackburn disappeared from the British aircraft industry's rolls, with so many other pioneering names. The government of the day decreed that the existing companies were too small and too numerous: 'shotgun weddings' were arranged, which created just two fixed-wing manufacturers – the British Aircraft Corporation and Hawker-Siddeley Limited, Blackburns being incorporated into the latter. In 1977, further rationalisation took place, with the creation of the monopolistic British Aerospace. Nevertheless, today, while famous centres such as Sir Geoffrey de Havilland's base at Hatfield and the old Sopwith works at Kingston-upon-Thames are gone completely, Brough, the birthplace of so many Blackburn designs of the early and middle years of this century, survives and flourishes as one of the major units of the nation's modern aerospace industry.

Right: In the first half of the nineteenth century, Capper Pass, from humble beginnings, founded his metal refining business in Bristol. His son, also named Capper Pass, developed what his father had begun into a thriving company.

Below: The decision to open a new plant on Humberside, at Melton, was taken in 1928, but the worldwide recession (the Slump) meant that the first blast furnace was not fired up until 1937. This photograph of the Melton works was taken in 1963. Today, the site is nothing but a desolation of weed-infested concrete and abandoned buildings.

EX · LIBRIS

ALFRED · CAPPER · PASS

After the Second World War, the Bean brothers started a market garden business on land between Brough and Welton, which became very much a family business. Over the years this enterprise has continued to prosper, as Humber Growers. A row of Beans: the members of the family and their immediate associates gather on a special occasion. They are, from left to right, standing: Philip, John Robert, Nicholas, Ian Sayer, Terry, Christopher, Eric, Robert, Alan Shirley, Roger Sayer, David. Seated: John, Olive, Willie.

Seven

The Humber –
Foreshore and Waterway

For the knowledge that men were sailing the Humber's waters over three thousand years ago, and in relatively sophisticated vessels, we are indebted to the work initiated and led by Edward Wright and his brother. Their discovery on Ferriby foreshore of substantial remains of five long boats have been scientifically dated as being from the Bronze Age – possibly as long ago as 1300 BC. Since those times, the broad-flowing Humber, with its shifting shoals and treacherous currents, has seen the comings and goings of many seafarers, as well as countless comings and goings in the transverse direction of various ferries which have linked the north and south banks, from at least Roman times, right up until the opening of the Humber bridge in 1981.

Brough Haven, which provided shelter for Humber Keels down the years.

Another glimpse of the foreshore at North Ferriby, facing east.

A tug towing a double line of lighters. Today, it is more likely that the view here would be of a container ship, out of Goole.

Still on Ferriby shore, but looking west, in about 1900, with a scene once familiar on the Humber. A hard-working tug with smoke issuing from its tall funnel can be seen and the sails of Humber Sloops are visible in the far distance. On the right is the old Ferriby brickyard with its landing stage while, beyond, the trees of the Long Plantation reach down to the waterside.

River Humber, Brough, E.Y.

Copyright

Raphael Tuck & Sons Ltd
London

Above: Both the date and the precise subject of this photograph are unknown. One can, however, say with some certainty that it is not a scene that one would be likely to see today. The sail of another Humber Sloop can be made out in the distance.

Left: In the winter of 1963-64, heavy snowfalls and very low temperatures ensured that it would be numbered amongst the coldest since the war. Even the Humber was frozen in places, as can be seen here, with two lightships frozen in at their moorings.

It was in 1937 that two young men, William and Edward Wright, foraging along the Ferriby shoreline, as they pursued their hobby of archeology, stumbled upon an ancient artefact buried in the Humber clay which was to lead to years of increasingly organised exploration of the site and to add significantly to the sum of our knowledge of early history. The discovery turned out to be a substantial portion of what was manifestly a very old boat ('Viking' was the thought which rather naturally first arose in the discoverers' minds). Energetic work ensued, while learned advice and help was sought and obtained. In due course, the first fragment illustrated above was identified as one end of the keel plank of a large vessel, predating the Vikings by a considerable time. These were early days for the Ferriby boats, while the science of carbon-dating was still in the future and it was to be some time before the significance of the finds would be appreciated to its full extent.

Above: By 1939, further portions of the boat, including parts of the side planks and more of the keel plank (above), had been brought to light, when more immediate history, in the shape of the Second World War, which temporarily transformed Ted Wright from amateur archeologist to tank commander in the East Riding Yeomanry, largely put a stop to further progress.

Left: With the end of hostilities, it was possible not only to resume work on a more organised basis, but also to begin to invoke the interest of the wider archeological establishment. Here the keel plank and neighbouring side planks of the first Ferriby boat (coded F1) can be seen, 27 August 1946.

Pieces of a second boat (F2) had been located during the war; these were brought out in 1946 and transported, together with F1, to the National Maritime Museum at Greenwich. Digging had by then, however, become only one side of the coin. From the outset, preservation presented a great problem, particularly given the limited knowledge of that subject then available.

The years passed; then, early in 1963, Ted Wright discovered the first traces of a third boat, F3. Without delay, plans were laid to excavate and recover these new fragments, whatever they might turn out to be. The above happy band of mudlarks was photographed engaged in that task in April 1963. From L to R: Bartlett, Naish, Mackey, Binns, Daae and Spalding, with the leader and original instigator, Ted Wright, facing them.

By now, the science of carbon-dating had been developed. In due course, suitable boat fragments were subjected to this analysis, which now appears to yield the conclusion, with a reasonable certainty, that the Ferriby boats date from around 1300 BC, within that period prehistorians have named the Bronze Age.

Left: A section of F2 keel plank exposed and awaiting the delicate task of removal from the grip of its imprisoning clay.

Below: The fragility of the ancient remnants represented a major concern. Here, F3 is standing on a plinth of clay, on which it will be supported while it is cautiously dragged from its long resting place.

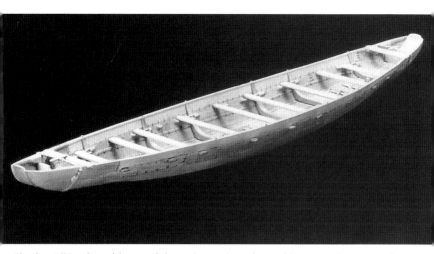

This fine 1/20 scale model was made by combining the evidence of the excavated remains and some reasonable hypothetical conclusions. With an assumed length of some 6m, the basic construction comprises a keel plank with side planks assembled edge-to-edge, held together by stitching with yew withies. The holes through which the withies were threaded are clearly to be seen in the excavated planks (see pages 123 and 124), while remnants of the withies themselves have also been found.

In recent years, the Ferriby boats have been joined by similar finds at Dover and by the Severn estuary. Thanks to the carbon-dating process, they have enabled us to learn that, over three thousand years ago, at least some of the inhabitants of these islands, far from simply paddling around in dug-out canoes, were building and sailing in vessels of far more sophisticated construction. No doubt, as the years pass, other significant discoveries will come to light, which will go to fill in other missing pieces in the jigsaw of our ancestral heritage.

Bibliography

History & Directory of East Yorkshire 1892 (Bulmer & Co.)
Bulman, David J. et al: North Ferriby – A Villagers' History
Neave, David (ed.): *South Cave* (Mr Pye Books)
Pearson, Richard: *Welton Sketch Book* (Hutton Press)
Wright, Edward: *The Ferriby Boats* (Routledge)

Acknowledgements

In a book which covers such a wide area, with many villages concerned, the sources of the nearly 200 photographs it contains have necessarily been many and various.

For the majority, I am indebted to Eileen Jones, Kay Laister, Mollie Cutts, Pam Walton, Ken Watson and Ted Wright, as well as to the East Yorkshire Local History Society (with the East Riding County Archives), the Brough Heritage Centre and Hull City Museums. Certain other photographs were contributed by Mrs Wiles, Pat Howlett, Anne Bousfield, Peter Greenfield, John Bean and Betty Allwood.

Apart from the photographs themselves, I have also been fortunate in the generous help I have been given, in large and small ways, in researching events and background, by all the above-named, as well as by Dick Chandler, Walter Blanchard, Wendy Dobbs, by Stan Field, Bob Ward and Steve Gillard of the Brough Heritage Centre, by Mrs C. Boddington of the East Riding County Archives and by Gail Foreman of Hull City Museums.